Synthetic & Renewable Fuels 2020

Topical Conference at the 2020 AIChE Annual Meeting

Online
16 – 20 November 2020

ISBN: 978-1-7138-2293-6

Printed from e-media with permission by:

Curran Associates, Inc.
57 Morehouse Lane
Red Hook, NY 12571

Some format issues inherent in the e-media version may also appear in this print version.

AIChE
120 Wall Street, FL 23
New York, NY 10005-4020

Phone: (800) 242-4363
Fax: (203) 775-5177

www.aiche.org

Additional copies of this publication are available from:

Curran Associates, Inc.
57 Morehouse Lane
Red Hook, NY 12571 USA
Phone: 845-758-0400
Fax: 845-758-2633
Email: curran@proceedings.com
Web: www.proceedings.com

TABLE OF CONTENTS

614585 Net-Zero Carbon Fuels Pathways Analysis

Abstract:

In 2019 the Net-Zero Carbon Fuels Technical Team (NZTT) was initiated by the U.S. DRIVE Partnership (a non-binding and voluntary government-industry partnership focused on advanced automotive and related energy infrastructure technology research and development) with the following Mission Statement: "Drive research, development, and demonstration of renewable energy solutions for the transportation sector through a consensus assessment of the carbon intensity, technoeconomic readiness, and challenges for volume implementation of net-zero carbon fuel pathways". The scope of NZTT includes investigation of liquid carbon fuels and proposing solutions for generating fuels with a reduced carbon intensity such that, from a life cycle carbon account standpoint, they have a net emissions profile of zero. As part of its scope, the Tech Team is conducting process analyses to examine the conditions required for economic viability and eventual demonstration of the most promising technologies. This ongoing analysis will be discussed in this presentation.

601277 Solar Fuel Production from CO_2 and H_2o Via the Hybrid CeO_2-CH_4 Redox Cyclic Process

Abstract:

Within the context of solar thermochemical fuel production, a redox process is discussed in which CeO_2 is chosen as the redox material due to its crystallographic stability and rapid kinetics. This material has been previously applied for the splitting of H_2O and CO_2 to produce syngas via a two-step thermochemical redox cycle using concentrated solar energy. In this solar cycle, temperature/pressure swings between the reduction and oxidation steps are thermodynamically favored for high fuel yield, but they impose severe thermal stresses on material structures and energy inefficiencies due to significant heat rejection. Alternatively, the CeO_2-CH_4 process is a hybrid solar-driven redox cycle that uses CH_4 to eliminate the temperature/pressure swings while further increasing the fuel output. This is because the addition of CH_4 as a reducing agent lowers the operating temperature by approximately 500°C, runs isothermally at about 1000°C, and achieves higher non-stoichiometry (δ) of $CeO_{2-\delta}$ thus indicating a higher proportion of oxygen exchange and, consequently, yielding a higher specific fuel output per unit mass of ceria. The hybrid CeO_2-CH_4 cyclic process is represented by two redox steps:

Solar endothermic reduction:

$CeO_2 + \delta CH_4 \leftrightarrow CeO_{2-\delta} + \delta CO + 2\delta H_2$ (1)

Non-solar exothermic oxidation:

$CeO_{2-\delta} + \delta CO_2 \leftrightarrow CeO_2 + \delta CO$ (2a)

$CeO_{2-\delta} + \delta H_2O \leftrightarrow CeO_2 + \delta H_2$ (2b)

The two redox reactions (i.e. eq. 1 and eq. 2) have opposing thermodynamic favorability with respect to temperature and are also affected by the extent of the reaction. Insights into the thermodynamic nature of the redox reactions allow one to predictably and efficiently control a reactor system by avoiding side reactions and operating in regimes of high reaction favorability.

Modelling and experimental studies are performed on a thermochemical reactor concept for effecting this hybrid solar-CH_4 redox cycle. The reactor concept for effecting the hybrid solar-CH_4 redox cycle consists of an alumina tube containing a porous structure of ceria exposed to an upward flow of the reacting gas, namely CH_4 during the reduction step (eq. 1) and CO_2/H_2O during the oxidation step (eq. 2). Various porous structures are investigated, including reticulated porous ceramic (RPC) foam-type structures and packed bed of pellets, granules, and other morphologies with varying specific surface areas (SSA). The tubular reactor is heated externally by a heat transfer fluid which carries high-temperature heat from the solar receiver. A lab-scale prototype reactor was fabricated using a 19 mm-diameter 300 mm-length Al_2O_3 tube, heated externally by an electrical furnace. Preliminary experimentation above 900°C indicated that an improved CeO_2 pelleted morphology increased the CH_4 conversion over that of the baseline RPC. Detailed experimental results will be presented.

A numerical heat and mass transfer model is developed to simulate the thermochemical reactor. OpenFOAM is applied to formulate and solve numerically the governing equations for fluid flow across porous media and combined convection-conduction-radiation heat transfer coupled to the chemical reactions. The effective transport properties of the RPC are incorporated [1]. The model is validated against laboratory experiments and applied for design optimization and scale-up.

Acknowledgements – This work was funded by the Swiss Federal Office of Energy (Grant No. **SI/501854-01**).

References

[1] S. Ackermann, M. Takacs, J. Scheffe, and A. Steinfeld, "Reticulated porous ceria undergoing thermochemical reduction with high-flux irradiation," International Journal of Heat and Mass Transfer, vol. 107, pp. 439-449, 2017.

591767 A Preliminary Study of Feasibility of Mechanical Vapor Recompression of Steam Carbon Dioxide Using Diaphragm Compressor for Energy Conservation in Carbon Capture

Abstract:

Energy consumption in carbon dioxide removal by using chemical solvents is the main point of concern due to the difference in operating conditions of the absorber and stripper besides the endothermic heat of desorption of chemical combined carbon dioxide in the stripper. The stripped gases leaving the stripper at about 105 C and 1 atm (g) contain uncondensed stripping steam to the extent of 50 vol%. The latent heat of the stripper outlet gases is normally rejected to cooling water. As an alternative , mechanical vapor recompression of the above gases to a high pressure around 30 bar (g) or more can facilitate recovery of thermal energy available in the steam by passing the compressed gases through the condensing coils of the reboiler to generate fresh stripping steam from the lean solvent. The main concern to be addressed in this method is the corrosiveness of the stripper outlet gases due to the presence of moisture. This can be taken care by preheating the gases to a temperature level above the dew point of the gas mixture and then compressing it . In the present work, a two stage metal diaphragm diaphragm compressor manufactured by Fluitron Inc, Pa, USA has been selected for study for pilot plant level investigation at IIT, Delhi. The compressor has a capacity to compress 1000 lpm of a 50% steam-carbon dioxide mixture at 105 C and 1 bar (g) to a pressure of 30 bar (g). From the preliminary results obtained in the present work, it appears feasible to compress a gas mixture containing steam and air or carbon dioxide by implementing appropriate preheating, temperature controls, materials of construction and safety measures. Tables 1-6 show some preliminary results obtained in the present work.

608303 Developing Tunable Nanoscale Hybrid Electrolytes for Combined CO2 Capture and Conversion

Abstract:

The electrochemical conversion of CO2 to chemicals and fuels using carbon free energy has recently been proposed as a pathway to achieving net carbon neutrality, due to the rapid cost reduction of renewable energies. Although there have been early successes in catalyst development and reactor design for CO2 reduction, it is also important to consider the design of electrolytes that can effectively deliver CO2 to electrodes with favorable product selectivity and reaction rates. Novel reactive materials are currently being employed as additives to aqueous and non-aqueous electrolytes in order to improve CO2 solubility, tune the selectivity of the CO2 reduction reaction and enhance reaction rates. In this talk, the current status of electrochemical CO2 conversion will be outlined through a brief discussion of relevant performance metrics, challenges and process requirements. The focus of this work is the development of Nanoparticle Organic Hybrid Materials (NOHMs)-based fluids as hybrid electrolyte systems that are capable of combined CO2 capture and conversion, without the need for a separate solvent regeneration step. Due to the complexity of these systems, a fundamental investigation of the structuring and organization of these nanoscale fluids is required. It has been observed that physical properties such as hydrodynamic size, viscosity and diffusion coefficient are highly dependent on the chemical nature and size of the secondary fluid molecule. Additionally, it was discovered that relevant transport properties of the untethered polymer were found to scale according to theoretical predictions. Interestingly, this behavior was not observed for NOHMs-based fluids. Lastly, the CO2 solubility of various NOHMs-based electrolytes has been measured and a change in the selectivity of the CO2 reduction reaction has been identified in the presence of these nanoscale electrolytes. These findings provide important insights into the design and development of a novel reactive medium that is capable of performing combined CO2 capture and conversion.

602024 Emissions Control Technologies for RTI's Non-Aqueous Solvent for Carbon Capture

Abstract:

Advanced water-lean solvents (WLS) for post-combustion CO_2 capture have been gaining interest due to their ability to reduce the parasitic penalty from energy needed for solvent regeneration. RTI International, with funding from the US Department of Energy, has been involved in development of its non-aqueous solvent (NAS), eCO_2Sol^{TM}. The solvent has demonstrated specific reboiler duties as low as 2.3 GJ/t-CO_2 at pilot plant scale CO_2 capture system at Tiller, Norway. Commercial implementation of these novel CO_2 capture technologies hinges on successful control of amine emissions. Vapor phase emissions can be controlled through use of water wash that scrub the amine components out of the treated gas as it exits the absorber column. While significant research work has been done to characterize and control aerosol-based emissions from aqueous solvents, similar information is not available for emissions from use of these novel water-lean solvents.

RTI has been conducting studies from fundamental and operational aspects to reduce the overall amine emissions from the WLS systems, specifically RTI's non-aqueous solvent eCO_2Sol^{TM}. The work has been carried out at 4-6 kW equivalent using the RTI's bench-scale gas absorption system (BsGAS). It consists of a CO_2 absorber column with intercoolers and water wash section, and a solvent regenerator column with a thermosiphon reboiler and inter-stage heaters. RTI's BsGAS was modified to add the aerosol generation system and monitoring equipment to determine the aerosol characteristics during the eCO_2Sol CO_2 capture process. The system is now capable of producing aerosols with the peak diameter and concentration of 50 mm and 1.2 $x10^7$ cm^{-3}, similar to those observed in the actual coal-fired power plant flue gas at the inlet of the CO_2 capture system.

RTI has conducted over 1,300 hours of parametric testing to evaluate the impact of the aerosols and operating conditions during the CO_2 capture with NAS on the overall amine emissions in the treated flue gas. A simulated flue gas, with 15% CO_2, 2.3-4.2% H_2O and balance air, was used in the study. SO_3 particulates were generated by reacting SO_2 with O_2 in air over a silica supported vanadium pentoxide (V_2O_5/SiO_2) catalyst at 450°C. The generated SO_3 particulates were injected into the simulated flue gas fed to the absorber. The SO_3 particulates mimic the presence of SO_X, fly ash and other particulates in the flue gas exiting the direct contact cooler and provides nucleation sites for growth of aerosols in the CO_2 capture process. A Scanning Mobility Particle Sizer and an Aerodynamic Particle Sizer were used to monitor the aerosol particles size and number distribution at various locations in the CO_2 capture system. The amine emissions at the absorber outlet and water wash outlet were monitored using a MKS MultiGas™ FTIR spectroscopy gas analyzer. Initial testing found that the CO_2 capture rate and water wash temperature were the two most dominant process factors on vapor emissions. Further testing fixed those two process variables to evaluate other variables including liquid/gas ratio (L/G, mass basis), intercooling profile (extent of intercooling at the top and bottom intercooler), temperature of the lean solvent to the absorber, the water wash L/G, temperature of the water wash return stream, and the inlet saturation temperature of the flue gas (which dictates both the extent of water in the flue gas, and the water wash temperature required to maintain water balance in the system). The CO_2 capture rate was maintained close to 90% capture rate.

Results from parametric testing suggested that the presence of the aerosols in the flue gas could increase the overall emissions by 10X compared to the baseline emissions from NAS's vapor pressure. The temperature difference between the temperature bulge seen in the absorber column and the water wash temperature impacts the particle growth, with the aerosol emissions increasing with the increase in the temperature difference. Most of the aerosols did not grow substantially in the system, and the particle concentrations remained nearly constant between the absorber inlet and wash outlet. Only a small portion of the particles were found to grow significantly. A second water-lean solvent was also tested and elicited insights into how to minimize vapor and aerosol emissions through solvent modification.

Results from parametric testing have established the emission baseline from water-lean solvent processes. The BsGAS has been modified to add in vapor and aerosol control technologies such as a water wash acidifier, an amine sorbent removal and regeneration system, high efficiency demisters, and a lime-coated filter which mimics a filter baghouse. Testing of the control technologies has shown their ability to significantly reduce the emissions from the water-lean solvents. Results from emission studies post-BsGAS modifications will be presented to show the effectiveness of the emission control strategies to minimize emissions from the CO_2 capture process using water-lean solvents.

606548 Optimization and Transport of CO_2 from Refineries in Southeast Texas for Offshore Subsea Storage

Abstract:

Behind fossil fuel-based power generation plants, petroleum Refineries are the second largest stationary source of CO_2emissions. Refineries often undergo process upgrades and plant expansions to maximize the crude throughput, thus maximizing profits. This in turn causes the amounts of CO_2 emissions from the refineries to vary annually, although predictably. For instance, process upgrades generally result in lower CO_2 emissions (~10%) based on total number of barrels of throughput. In addition, CO_2 emissions increase by 85 – 90% when the plant capacity is doubled. Due to these constant changes in the refineries, estimating the amount of carbon dioxide to predict the size of the compressor and the recompression station required for the purposes of Carbon Capture and Storage (CCS) becomes complicated.

In this study, the Petroleum Refineries present in the Beaumont and Port Arthur, Texas area were analyzed for CO_2output using data available from EPA's Facility Level Information on GreenHouse Gases Tool (FLIGHT). Refinery Capacity and actual production was determined from quarterly and annual reports from the respective refineries. The petroleum refineries considered for this study include Valero Refinery (Port Arthur, TX), Total Petrochemicals & Refining USA, Inc. (Port Arthur, TX), ExxonMobil Beaumont Refinery (Beaumont, TX), and, Motiva Enterprises, LLC (Port Arthur, TX).

Among the six different types of CO_2 emission sources found within refineries, Stationary Fuel Combustion Sources, Catalytic Cracking and Reforming Units, and Sulfur Recovery Units accounted for about 80 – 85% of the total carbon dioxide emissions. The CO_2 emissions produced from these units was found to have the highest correlation with crude throughput. These findings were used to size compressors and a design CO_2 gathering network for delivery of $CO_2$from these facilities to offshore, subsea locations in the Gulf of Mexico.

607589 Dynamic Modeling, Optimization, and Control Studies of a Moving Bed Process for CO_2 Capture Using a Micro-Encapsulated Solvent

Abstract:

A number of potential solvents for CO_2 capture exhibit undesirable characteristics like high viscosity, solids formation thus making it difficult to use them in conventional towers [1]. Micro encapsulation of carbon sorbents (MECS) can help in overcoming these issues seen in highly viscous and phase change materials. Micro-encapsulation of solvents is a technology proposed [2] for post combustion carbon capture where the solvent of interest is encapsulated inside a polymer. Microencapsulation of carbon sorbents (MECS) involves fabrication of microcapsules containing a solvent with CO_2 capture capabilities. Even though capsules are manufactured with a known solvent concentration, water is known to permeate some capsule materials. H_2O gain or loss due to diffusion into or out of the microcapsules can significantly affect the solvent concentration and is one of the key operational challenges for reliable and efficient operation of MECS technology. While there are a few studies [2-3] that have investigated different reactor configurations for MECS, no studies have been conducted with the intent to understand and mitigate the issue of time-varying water concentration. Extracting solvent from microcapsules to measure the concentration may require collecting huge number of microcapsules from the reactor. The ambient conditions can further change the water content and may not represent the true concentration value in the reactor. Therefore one must rely on other measurable process variables to get an estimate of capsule concentration.

The present work is focused on dynamic modeling, optimization, and control of a moving bed reactor for MECS, where controlling the desired solvent concentration is one of the key control objectives. Moving bed reactors have a strong potential for solid-sorbent based CO_2 capture [4-5]. The counter-current contact between gas and capsules can help to achieve larger CO_2 driving forces in the absorber and regenerator and so is examined in this work. Also, the heat recovery from the capsules leaving the regenerator can help in pre-heating the capsules flowing from absorber to regenerator. Overall, this can reduce the regeneration energy requirements of the process. A 1-D multi-scale moving bed absorber and regenerator using a pre-existing microcapsule model [4] is developed in Aspen Custom Modeler (ACM) for MECS using Na_2CO_3 as the encapsulated solvent. The capsule level model used in the moving bed reactor is validated with experimental data in our previous study [4]. A steady-state optimization of the moving bed configuration that minimizes the Equivalent Annual Operating Cost (EAOC) by considering key operating and design variables is performed. Traditional and model-based advanced controllers are developed for controlling the capture percentage, and solvent concentration within the capsule in the face of input disturbances. An ARMAX type model based on information about measured variables from the rigorous process model will be exploited to estimate solvent concentration and to provide satisfactory control performance.

References

1. Stolaroff, J. K. et al., Microencapsulation of advanced solvents for carbon capture. Faraday Discussions. DOI: 10.1039/c6fd00049e (2016).
2. Vericella, J. J. et al. "Encapsulated liquid sorbents for carbon dioxide capture", Nature Communications, 6:6124, 1-7, 2015.

3. Hornbostel, et al, Packed and fluidized bed absorber modeling for carbon capture with micro-encapsulated sodium carbonate solution, Applied Energy 1192-1204, 2019.
4. Kotamreddy, et al., Process Modeling and Techno-Economic Analysis of a CO_2 Capture Process Using Fixed Bed Reactors with a Microencapsulated Solvent, Energy & Fuels 2019 33 (8), 7534-7549 DOI: 10.1021/acs.energyfuels.9b01255.
5. Knaebel, K.S., Temperature swing adsorption system. 2009, Google Patents.
6. Mondino, Giorgia et al. "Effect of Gas Recycling on the Performance of a Moving Bed Temperature-Swing (MBTSA) Process for CO_2 Capture in a Coal Fired Power Plant Context." Energies, 10(6), 745. 2017.

608069 Enhanced CO_2 Uptake and Selectivity upon Incorporation of Task-Specific Ionic Liquids into Highly Porous MOF-177 at Post-Combustion Capture Condition

Abstract:

Recently, Metal-organic-frameworks (MOFs) have gained tremendous attraction for application in CO_2 capture and storage due to an exceptional surface area, pore volume and for their tunability. Although some of the MOFs have shown excellent CO_2 uptake at high pressure, lower uptake at post combustion condition (partial pressure of CO_2 is 0.15 bar) is a major hindrance for widespread application in CO_2 capture operations. Researchers have deployed various strategies to enhance CO_2 uptake such as grafting amines functionalities covalently to the ligands, impregnating amines or ionic liquids (ILs). Post-synthesis modification of pores of solid sorbents by impregnation of amino functionalized ionic liquid also known as task-specific ionic liquid (TSILs) to enhance CO_2 capture capacity has shown promising results.

In this study, 1-ethyl-3-methylimidazolium [Emim] cations with Glycine [Gly] and Alanine [Ala] as reactive Amino Acid (AA) anions forming [Emim][Gly] and [Emim][Ala] as TSILs were immobilized into the pores of MOF-177 through the incipient wetness technique. TSIL@MOF-177 composites sorbents were prepared for three different loadings (10, 20 and 30) wt.% of TSIL and the developed composites were characterized and investigated from the engineering perspective in terms of CO_2 capture capacity, selectivity, and enthalpy of adsorption. CO_2 adsorption isotherms were measured at three different temperatures (303, 313 and 323 K) in the pressure range varying from (0.1 to 10) bar. In addition, N_2 adsorption isotherms were also measured for all composites and the pristine MOF-177 from which CO_2/N_2 ideal selectivities were calculated. In addition, experimental adsorption isotherms were modelled using the Freundlich model and the isosteric heats of adsorption were calculated and compared to those of the pristine MOF-177.

Impregnation of TSILs into the highly porous metal organic frameworks (MOF-177) had a remarkable impact on CO_2 capture and selectivity. The highest CO_2 uptake achieved by [Emim][Gly]@MOF-177 composite was 0.45 mmol/g-solid and 0.42 mmol/g-solid at 0.2 bar, 303 K for [Emim][Ala]@MOF-177 composite which were 3 and 2.8 times higher than the pristine MOF-177 at the same conditions, respectively. It was also observed that the optimum loading of TSIL was 20 wt.% as further increase to a loading to 30 wt.% did not results in any incremental CO_2 uptake. CO_2 selectivity over N_2 was also enhanced from 5 for the pristine MOF-177 to 13 and 11 for [Emim][Gly] and [Emim][Ala], respectively, at 0.2 bar and 313 K. This significant improvement of both CO_2 uptake and selectivity can be attributed to the chemisorption interaction between CO_2 and -NH_2 functional group present in the amino acid anion in TSIL which was also reflected in the increase in the isosteric heat of adsorption (Q_{st}) values from the value of the pristine MOF. Significantly high CO_2 uptake and CO_2/N_2 selectivity at the low pressure region (<1.0 bar) makes these functionalized MOFs promising candidates for the post-combustion CO_2 capture operation.

600872 Convert CO_2 to Lycopene Directly in Water Splitting–Biosynthetic System Via Renewable Energy

Abstract:

CO_2 conversion has been intensively studied in recent years for global environmental and energy issues. Considerable high-performance electrocatalysts have been reported to catalyze CO_2 reduction reactions (CRRs) into different chemicals with renewable electricity. However, CO_2-based C–C coupling reductive reactions with high efficiency and selectivity have not been realized yet. Based on inorganic catalysis, CO and formate are usually obtained from CRRs. Recently, water splitting-biosynthetic hybrid system has been proved to be a promising platform for CO_2 reduction to complex compounds such as isopropanol, terpenes and poly(3-hydroxybutyrate). As a combination of inorganic water splitting system and H_2-oxidizing autotrophic microorganisms, microbial can take energy from H_2 to convert CO_2 to different chemicals, biofuels and medical intermediates with high selectivity. Besides, cellular metabolism can be rewired to the formation of the products of interest further via synthetic biology approaches. In this study, *Ralstonia eutropha* C5/pBBR1-crtEBI, a recombinant strain that can convert CO_2 to lycopene under autotrophic conditions, was constructed. Then a microbial electrosynthesis (MES) system was established by coupling C5/pBBR1-crtEBI with water splitting system. At a constant current of 18 mA, C5/pBBR1-crtEBI could grow in this system and OD600 increased from 0.3 to 1.17 in 8 days. After systematic optimization of the MES system, the maximum lycopene production reached 1.68 mg/L while no lycopene was detected in the control strain. This system also showed excellent stability, where lycopene can be synthesized continuously for more than 8 days. To the best of our knowledge, for the first time, lycopene can be successfully synthesized in water splitting–biosynthetic system driven by renewable electricity. This work provides a new idea for CO_2 conversion to complex products simultaneously with storage of renewable energy.

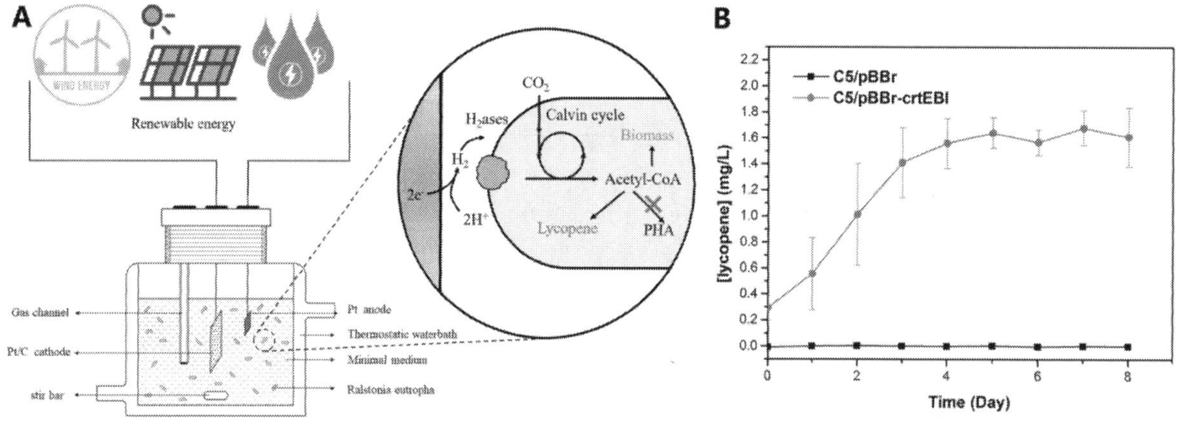

598424 Techno-Economic and Life Cycle Analysis of Dimethyl Ether Production Via the Bi-Reforming Pathway for Transportation Fuel

Abstract:

Dimethyl-ether (DME) is a diesel fuel-substitute commercially produced via Natural Gas (NG) reforming. This approach is economical but generates significant greenhouse gas (GHG) emissions. This study investigates the Techno-economic (TEA) and Life Cycle Assessment (LCA) of DME production via bi-reforming of waste CO_2 streams and methanol synthesis and dehydration.

A conceptual DME production plant with a 500 metric tonne per day capacity is simulated in Aspen Plus^{TM}. In the model, CO_2 is sourced from either an Ammonia production facility (CFA) or from Landfill gas (CFL). The TEA evaluates the DME minimum fuel-selling price (MFSP) over a 30-year project lifetime with a 12% internal rate of return. The LCA estimates the well-to-pump GHG emissions using GREET.net over a 100-year horizon. TEA and LCA results are compared to conventional diesel production. Sensitivity analysis of the MFSP to key TEA parameters including DME production, capital cost, and CO_2 price is investigated.

The estimated minimum fuel-selling price (MFSP) is \$1.03 and \$1.05/gal for the CFA and CFL cases, respectively. The equivalent diesel price ranges from \$1.93 to \$1.96/gal, which would be competitive with the market price of conventional diesel fuel. The well-to-pump lifecycle analysis shows that emissions range from -38.2 to 34.8 kg CO_2 eq/MMBtu. Using landfill gas sourced CO_2, the process achieves negative emissions by avoiding the release of more potent GHGs whereas using CO_2 from an ammonia plant results in higher emissions than conventional diesel (17.7 kg CO_2 eq/MMBtu). These results indicate that bi-reforming could be an economical and sustainable pathway for DME production as a diesel substitute.

593952 Photocatalytic Conversion of Carbon Dioxide to Methanol Using Visible Light Responsive MOF Nanocomposites

Abstract:

Photocatalytic conversion of carbon dioxide to methanol using visible light responsive MOF nanocomposites

Due to global warming and shortage of energy, the demand for sustainable energy sources has been increasing. Carbon emissions are significant contributors to the problem of global warming. This opportunity can be targeted incorporating unique compounds like photo catalysts or membranes of metal-organic frameworks (MOFs). MOFs application in fields like energy storage, heterogeneous catalysis, sensing and gas storage is found to be very promising. The objective of this project is to successfully synthesize and characterize the bismuth , graphene and copper based MOF nanocomposites for utilization of visible light for photocatalytic reduction of carbon dioxide to renewable fuels like methanol and to compare their efficiency in terms of methanol production as it will help us to determine the better conditions for the conversion of CO_2.The MOF nano-composites synthesised and characterized were used for the conversion of carbon dioxide to methanol are Cu_3BTC_2, rGO-CuO nanocomposites and ZIF-8/$BiVO_4$. In this work, three different nanocomposites were synthesized successfully using three different approaches. Cu_3BTC_2 metal organic framework was synthesized using a microwave method whereas rGO-CuO nanocomposite was prepared by combining rGO and CuO under a nitrogen atmosphere with the help of APTMS. The ZIF-8/$BiVO_4$ nanocomposite was synthesized via in-situ growth method where ZIF-8 nanoparticles were made to grow on $BiVO_4$ sheets, $BiVO_4$ for this nanocomposite was synthesized via a hydrothermal method. Synthesis of Cu_3BTC_2 is carried out by dissolving aqueous Copper (II) nitrate trihydrate solution in a solution of 1,3,5-benzene tricarboxylic acid (H_3BTC) in 1:1:1 mixture of ethanol, N,N-dimethylformamide (DMF) and DI water. rGO-CuO nanocomposites are synthesized using in-situ growth method of implanting CuO nanorods into the rGO framework. ZIF-8/$BiVO_4$ framework is obtained by in-situ growth of ZIF-8 on $BiVO_4$ nanosheets. The MOFs are then dissolved into a solution and CO_2 is passed through them under a visible light irradiation for conversion to methanol. The corresponding yields obtained through each MOF are compared and analysed for maximum output. Characterization was done using XRD analysis, the surface area of the sample was determined by BET and the morphology was determined by STEM-EDS analysis and Raman spectroscopy and the data obtained indicates successful synthesis of the sample and also sheds light on some of the internal structure properties of the sample. The crystallinity and the morphology of rGO-CuO nanocomposites were determined by XRD and STEM analysis respectively and the same holds true for ZIF-8/$BiVO_4$ nanocomposites. Thus this newly designed photoreactor can yield competitive amounts of fuels from CO_2 photoreduction via visible light/solar light irradiation using synthesized MOF nanocomposites.

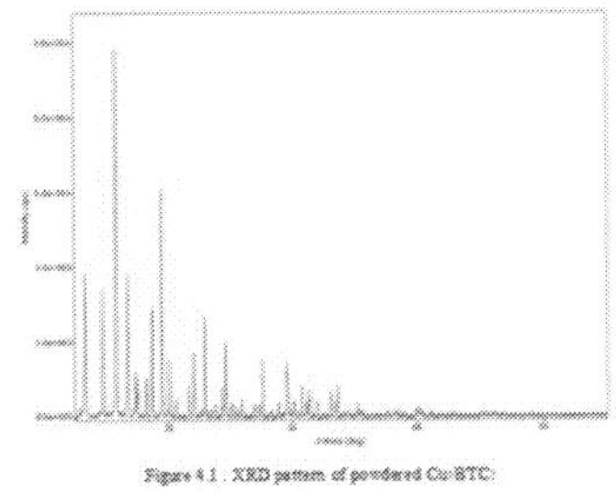

Figure 4.1 : XRD pattern of powdered Cu-BTC:

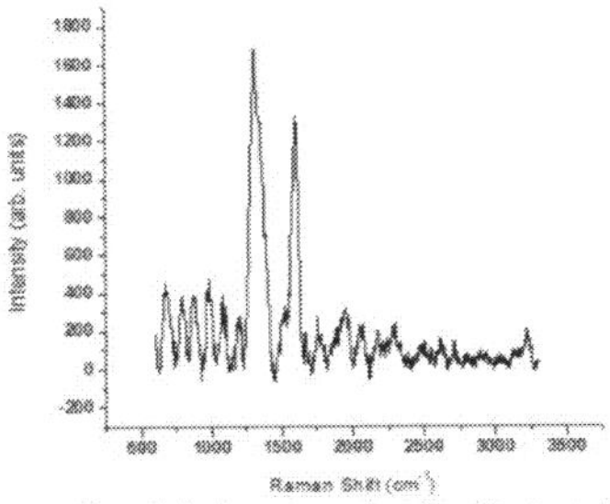

Figure 4.6 : Raman spectra of synthesized rGO

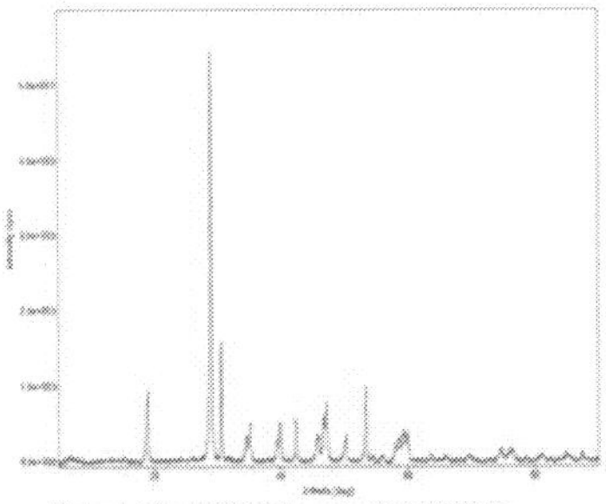

Figure 4.8 : XRD pattern for ZIF-8/$BiVO_4$.

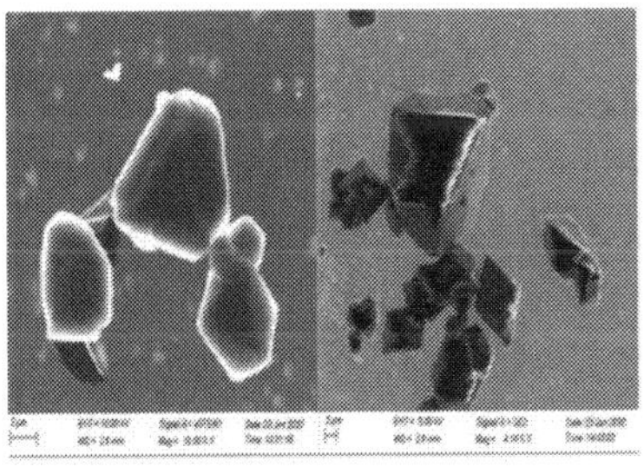

Figure 4.3 : STEM images of Cu_3BTC_2

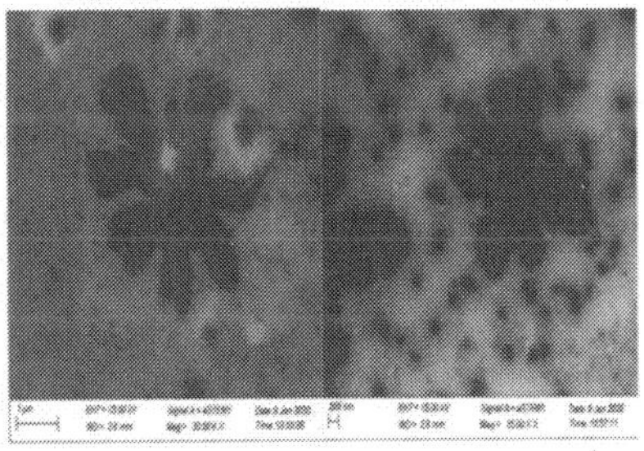

Figure 4.7 : STEM images of rGO-CuO

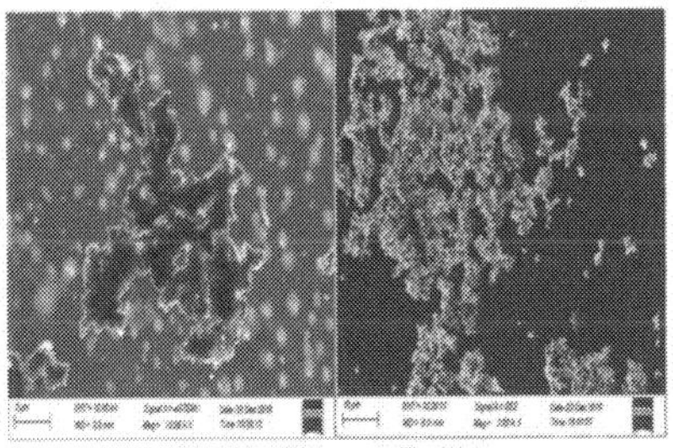

Figure 4.9 : STEM images of ZIF-8/$BiVO_4$

614569 The Important Role of Hydrogen in Synthetic Fuels

Abstract:

Synthetically producing energy-rich chemicals from diverse and abundant domestic resources offers the promise of providing a plentiful supply of sustainable fuels to meet future U.S. energy needs. Opportunities for producing strategically-important synthetic fuels include hydrogen from water, hydrocarbon fuels from carbon dioxide and hydrogen/water, and ammonia from di-nitrogen and hydrogen/water. Most fuels are currently produced from fossil resources using energy-intense high temperature processes, but advanced processes for fuel synthesis utilizing various sustainable energy resources in conjunction with air and water can provide critical supplements to help meet near- and longer-term energy demands. Most pathways for synthetic fuels rely heavily on the availability of an abundant and sustainable supply of hydrogen. More generally, synthetic fuels production is an important end-use sector supported by the U.S. Department of Energy's (DOE) H2@scale initiative. Specific challenges and opportunities for a new generation of synthetic fuels production will be discussed in the context of H2@Scale as well as other DOE efforts including the HydroGEN Energy Materials Network Consortium on Advanced Water Splitting Materials.

609969 Metal Oxide Based Solar Driven Thermochemical Water Splitting Cycle: Solar Reactor Efficiency Analysis

Abstract:

The conversion of solar energy into chemical energy in the form of alternative fuel such as H2 provides a promising option for sustainable future energy pathways. Hydrogen can be produced with solar thermal energy via water splitting reactions. Hydrogen used in this manner is an energy carrier. Its advantages include a high energy density (on a mass basis) and non-polluting nature. The metal oxide (MO) based two-step solar thermochemical cycle is a potentially advantageous way to produce hydrogen via water splitting. In the past, several MO-based redox systems were theoretically and experimentally studied as potential thermochemical water splitting reactions. These include zinc oxide cycles, iron oxide cycles, tin oxide cycles, terbium oxide cycle, mixed ferrite cycles, ceria cycles, and perovskite cycles. In case of zinc and tin oxide cycles, due to the volatile nature of the MOs involved, the loss of reactive material due to high-temperature operation is inevitable. In the case of the iron oxide, ferrite, ceria, and perovskite cycles, the complete reduction of the MOs is not possible, and hence lower amount of H2 production is expected. Therefore, it is highly essential to investigate new thermochemical cycles for H2 production via thermochemical water splitting. In this paper, the solar reactor efficiency analysis of metal oxide-based two-step solar thermochemical water splitting cycle was investigated in detail. Computational thermodynamic modeling of this cycle is performed (in two sections) using commercial thermodynamic HSC Chemistry software and databases. In section one, the thermodynamic equilibrium compositions during the solar thermal reduction and oxidation via water splitting were determined. In section two, an energy and exergy analysis of the metal oxide cycle was carried out. The solar reactor absorption efficiency, the required solar energy input, radiation heat losses, the rate of heat rejected by the quench unit and the water-splitting reactor, and the solar-to-fuel conversion efficiency of the metal oxide cycle were determined via computational thermodynamic modeling and the results reported.

593068 Estimation of Electronic Entropy Contributions to Redox Thermodynamics in Ceria

Abstract:

The primary motivation of this work is to develop a method through which the entropy change of oxygen vacancy formation may be computed from first principles for strongly correlated metal oxides which exhibit intrinsic spin polarization. Namely, a methodology via which the electronic structure may be utilized to predict electronic entropy contributions to the overall free energy change. Leveraging electronic structure properties facilitates the process by circumventing the need for extensive combinatorics and statistical methods which are conventional approaches for entropy calculations within the realm of statistical physics. Nonstoichiometric cerium dioxide (CeO2) and the manganite perovskites (AMnO3) form a prototypical composition space meeting the criteria above and
possessing a high degree of versatility with applications ranging from electrocatalysis and solar thermochemical energy storage to electronic and magnetic devices. The acquisition of a fundamental understanding of the mechanisms governing entropy in these materials would enhance the rate at which
they may be optimized and tailored for current applications and potentially enable the discovery of novel applications for which they are ideal candidates.

It is commonly assumed that finite temperature electronic contributions to the free energy of insulators and semiconductors may be neglected due to the lack or absence of electron density near the
Fermi level. This may be the case when treating ideal, pristine systems since there are no electronic levels available for population due to thermal excitation of electrons near the valence band maximum. This sizable bandgap in medium to wide gap materials limit the number of electronic configurations accessible to the highest occupied levels. The validity of this assumption is highly questionable however, especially when charged defect levels are present within the energy gap increasing the density
of states near the Fermi level. Shallow defect levels with unpaired electrons near the VBM would allow
for an increase in the number of microstates for thermal population of electrons contributing to the electronic configurational free energy. The electronic density-of-states data from a structural relaxation
at the HSE06 level of theory with a 12-atom Ce4O8 bulk supercell was implemented to calculate the Fermi level (electron chemical potential) as a function of temperature. The Fermi-Dirac distribution along with the normalized DOS were utilized to estimate the majority carrier concentration (assuming only electronic charge carriers) via the difference between the conduction band and valence band integrals.

A greedy algorithm is utilized to evaluate the Fermi level at which the error between the calculated carrier concentration (via above parameters) at a given temperature and the predicted/estimated carrier concentration is minimized. The corresponding electronic entropy agrees relatively well with results from work by Naghavi et al. in magnitude and variation with respect to temperature. Further validating the significance of electronic entropy contributions to the energetics of oxygen vacancy formation in ceria and other similar mixed conducting oxides which may be comparable to if not greater than structural configurational and vibrational entropic terms. Recent work by Lany highlighted this relationship between temperature, Fermi level and charged defect formation energies specifying that electronic entropy is the physical or thermodynamic consequence of this phenomenon. Specifically, the energy level of the oxygen vacancy, the conduction band effective mass and the temperature dependence of the conduction band minimum were claimed to be the electronic structure properties that govern the electronic entropy contribution to vacancy formation free energies.

The work presented herein provides numerical evidence to support the claims made by Lany in a purely analytical deduction.

1. Naghavi, S. Shahab, et al. "Giant Onsite Electronic Entropy Enhances the Performance of Ceria for Water Splitting." Nature Communications, vol. 8, no. 1, 2017, doi:10.1038/s41467-017-00381-2.
2. Lany, Stephan. "Communication: The Electronic Entropy of Charged Defect Formation and Its Impact on Thermochemical Redox Cycles." The Journal of Chemical Physics, vol. 148, no. 7, 2018, p. 071101., doi:10.1063/1.5022176.
3. Kolodiazhnyi, Taras, et al. "Disentangling Small-Polaron and Anderson-Localization Effects in Ceria: Combined Experimental and First-Principles Study." Physical Review B, American Physical Society, 23 Jan. 2019, journals.aps.org/prb/abstract/10.1103/PhysRevB.99.035144.
4. Sun, Lu, et al. "Disentangling the Role of Small Polarons and Oxygen Vacancies in \$\Mathrm{Ce}{\ Mathrm{O}}_{2}\$." Physical Review B, American Physical Society, 1 June 2017, journals.aps.org/prb/ abstract/10.1103/PhysRevB.95.245101.

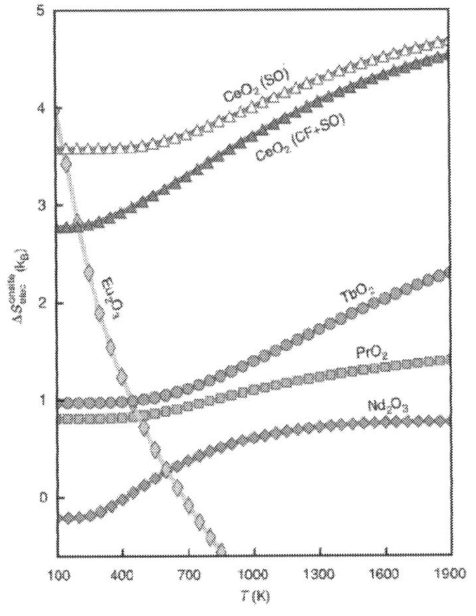

Fig. 2 Calculated ΔS^{onsite}_{elec} for lanthanides ions. Predicted electronic entropy of reduction per oxygen vacancy for the lanthanide oxides studied in this work. At high temperature, reduction of CeO_2 has the highest ΔS^{onsite}_{elec} followed by reduction of TbO_2 (see Table 1)

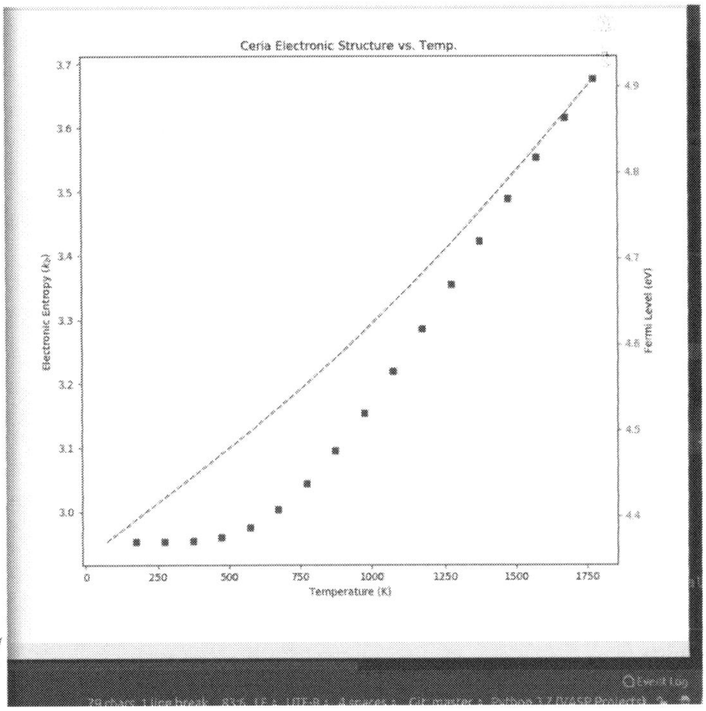

600480 On-Sun Hydrogen Generation from Photoelectrochemical Water-Splitting Devices

Abstract:

Photoelectrochemical (PEC) water-splitting is a promising technology that uses sunlight to directly split water into hydrogen and oxygen, providing a storable form of chemical fuel.[1] While several semiconductor systems have shown the capability for unassisted solar hydrogen production, III-V semiconductors have demonstrated the highest solar-to-hydrogen (STH) conversion efficiencies and best device longevity.[2–4] However, to date, work in the field has focused almost exclusively on laboratory conditions, with very few PEC systems having been tested outdoors and on-sun.

In this work, we present unassisted PEC hydrogen generation under natural solar illumination using a tandem III-V semiconductor absorber device paired with a non-precious metal HER catalyst. An on-sun photoreactor platform was designed for PEC material testing and benchmarking that incorporates a 2-axis solar tracker to maximize sunlight utilization and a gas collection system.[5] Nearby insolation and environmental monitoring instruments at the NREL Solar Radiation Research Laboratory provide a full record of the solar flux and environmental conditions that allow for more accurate comparisons of on-sun and in-lab performance. Tandem, lattice-matched $GaInP_2$/GaAs absorbers (1.8/1.4 eV bandgaps) that have been demonstrated with a high quality fabrication and yield >10 % STH efficiency[2,3] are coated with a MoS_2 HER catalytic and protective layer. Photoelectrochemical performance was characterized by current-voltage and chronoamperometric measurements under both simulated and natural sunlight illumination, while material structure is probed by XPS and SEM.

The MoS_2-protected tandem absorbers exhibit >10% STH efficiency under both laboratory and on-sun conditions. Over the course of a day of PEC testing under natural sunlight, >11 standard mL of hydrogen is generated with no external bias. We will discuss catalyst and semiconductor stability and cell design challenges that limit PEC durability and hydrogen yield.

(1) Seitz, L. C.; Chen, Z.; Forman, A. J.; Pinaud, B. A.; Benck, J. D.; Jaramillo, T. F. Modeling Practical Performance Limits of Photoelectrochemical Water Splitting Based on the Current State of Materials Research. *ChemSusChem* **2014**, *7* (5), 1372–1385.

(2) Young, J. L.; Steiner, M. A.; Döscher, H.; France, R. M.; Turner, J. A.; Deutsch, T. G. Direct Solar-to-Hydrogen Conversion via Inverted Metamorphic Multi-Junction Semiconductor Architectures. *Nat. Energy* **2017**, *2* (4), 17028.

(3) Khaselev, O.; Turner, J. A. A Monolithic Photovoltaic-Photoelectrochemical Device for Hydrogen Production via Water Splitting. *Science* **1998**, *280* (5362), 425–427.

(4) Cheng, W.-H.; Richter, M. H.; May, M. M.; Ohlmann, J.; Lackner, D.; Dimroth, F.; Hannappel, T.; Atwater, H. A.; Lewerenz, H.-J. Monolithic Photoelectrochemical Device for Direct Water Splitting with 19% Efficiency. *ACS Energy Lett.* **2018**, *3* (8), 1795–1800.

(5) HydroGEN Consortium. On-Sun Photoelectrochemical Solar-to-Hydrogen Benchmarking https://www.h2awsm.org/capabilities/sun-photoelectrochemical-solar-hydrogen-benchmarking.

AUTHOR INDEX